Genetic Gastronomy

Cultivating Nutritious Food Using Genetically Modified Crops

Jona Morningstar

Table of Contents

1 Introduction to Genetic Gastronomy

Genetic modification offers a powerful toolkit for enhancing the nutritional profile of staple crops. Imagine rice enriched with beta-carotene, effectively combating vitamin A deficiency in millions. Or consider golden rice, a testament to the potential of genetic engineering to address critical public health issues. This targeted approach contrasts sharply with traditional fortification methods, often less efficient and facing challenges in consistent nutrient delivery. Biofortification, in contrast, allows for the integration of essential nutrients directly into the plant's genetic makeup, resulting in a more natural and readily available source of vital compounds.

The development of pest-resistant crops represents a significant advancement in sustainable agriculture. By incorporating genes that confer resistance to specific pests, we can drastically reduce or even eliminate the need for harmful chemical pesticides. This not only minimizes environmental damage but also safeguards human health, reducing exposure to pesticide residues in food. The economic benefits are also substantial, lowering costs for farmers and consumers alike. Furthermore, the reduced need for pesticide applications translates to a more environmentally friendly agricultural system. This holistic approach demonstrates the power of genetic engineering to address interconnected challenges in food production and environmental sustainability.

Beyond pest resistance, genetic modification has been instrumental in developing crops that can withstand harsh environmental conditions. Drought-tolerant varieties are particularly crucial in arid and semi-arid regions, where water scarcity poses a significant constraint on agricultural productivity. These resilient crops allow farmers to maintain yields even during periods of low rainfall, ensuring food security in vulnerable areas. Moreover, the development of disease-resistant crops protects against significant yield losses caused by pathogens and pests, reinforcing food stability during adverse weather events. Such advancements are especially pertinent given the increasing frequency and intensity of extreme weather patterns linked to climate change.

The ethical considerations surrounding genetically modified organisms (GMOs) are complex and multifaceted. Concerns about potential environmental impacts, such as the development of herbicide-resistant weeds, necessitate careful monitoring and risk assessment. Public perception of GMOs is also a crucial factor, often shaped by misinformation and a lack of understanding. Transparency in research and clear communication about the benefits and risks of GMO technology are essential to fostering public trust and ensuring the responsible deployment of these powerful tools. The ongoing dialogue surrounding these issues is critical for shaping policies and practices that maximize the benefits of genetic modification while mitigating potential downsides.

Addressing global food security is a paramount challenge in the 21st century, and genetic gastronomy plays a significant role in providing viable solutions. Increased crop yields through genetic modification offer a crucial means of addressing food shortages, especially in regions facing population growth and environmental stresses. However, simply increasing production is insufficient; equitable access to food remains a critical concern. Strategies must also focus on improving food distribution networks and addressing socio-economic factors that hinder food availability and affordability for vulnerable populations. A comprehensive approach is required, integrating genetic modification with sustainable farming practices and robust social policies, to create a truly food-secure world. The potential of genetic gastronomy is vast, but realizing its full benefits demands a careful consideration of the various social, environmental, and economic dimensions involved.

The future of genetic gastronomy hinges on continuous innovation and responsible implementation. Emerging technologies, such as advanced gene editing tools, hold tremendous promise for further enhancing the nutritional content, resilience, and yield of crops. However, the regulatory landscape must keep pace with these advancements, ensuring the safe and ethical use of new technologies. Furthermore, fostering public dialogue and education is crucial for dispelling misconceptions and fostering acceptance of genetically modified crops as a vital component of sustainable food systems. The journey toward a food-secure future demands not only technological advancements but also a concerted effort to address social, economic, and environmental considerations. This collaborative approach is vital to unlocking the full potential of genetic gastronomy.

1.1 The Promise of GMOs

Genetically modified organisms, or GMOs, represent a powerful tool in our arsenal against global food insecurity. Their potential to enhance crop yields and nutritional value is immense, offering a pathway towards a more sustainable and food-secure future. This isn't about replacing traditional farming methods entirely, but rather augmenting them with precision technologies that address specific challenges. We're talking about crops that can thrive in harsher conditions, requiring less water and fewer pesticides, ultimately leading to lower production costs and increased accessibility for consumers worldwide.

The benefits extend beyond simply increasing food production. Imagine crops fortified with essential vitamins and minerals, combating widespread malnutrition, particularly in developing nations. Genetic modification allows for targeted enhancements, addressing specific nutritional deficiencies within populations. This isn't science fiction; biofortification programs already demonstrate the real-world impact of this technology, improving the lives of millions. The potential for tailored nutrition, designed to meet the needs of specific demographics, is truly transformative. This means a future where food isn't just sustenance but a powerful tool for public health.

Consider the environmental implications. Pest-resistant crops significantly reduce the need for synthetic pesticides, minimizing their negative impacts on ecosystems and human health. Similarly, drought-tolerant varieties conserve precious water resources, making agriculture more sustainable in arid and semi-arid regions. These advancements directly address some of the most pressing environmental concerns facing our planet, presenting a more environmentally conscious approach to feeding a growing global population. The reduction in pesticide use, in particular, has substantial benefits for biodiversity and human health. Reduced pesticide runoff means cleaner waterways and healthier ecosystems.

However, the adoption of GMOs isn't without its challenges. Consumer perception and regulatory hurdles remain significant obstacles. Addressing these concerns requires transparency, rigorous scientific research, and open public dialogue. We need to move past unfounded fears and engage in informed discussions about the risks and benefits. Education is crucial here, empowering consumers with the knowledge to make informed choices about the food they consume. This isn't just about providing scientific evidence; it's about fostering trust and building a shared understanding of the technology's potential and limitations.

Furthermore, equitable access to GM technology is paramount.

We must avoid a scenario where these advancements only benefit wealthy nations, leaving developing countries behind. International collaborations and support are necessary to ensure that the benefits of GMOs are shared globally, contributing to a more just and sustainable food system. It's about fostering innovation that is inclusive and accessible to all, not just a privileged few. The technology should serve humanity as a whole, not just a select group.

The promise of GMOs lies not just in their potential to increase food production but in their capacity to improve nutrition, enhance environmental sustainability, and contribute to global food security. Addressing the challenges surrounding consumer perception and equitable access is crucial to realizing this promise. With thoughtful regulation, transparent communication, and continued scientific innovation, GMOs can play a critical role in building a more resilient and sustainable food system for all. It requires a collaborative effort from scientists, policymakers, and the public to harness the power of this technology responsibly, ensuring its benefits are shared widely and its risks are mitigated effectively. The path forward demands a multifaceted approach, combining scientific advancements with ethical considerations and robust regulatory frameworks. Only then can we unlock the true potential of GMOs and their transformative power. The future of food security might well depend on our ability to embrace this technology, responsibly and effectively.

1.2 Ethical Considerations

The development of genetically modified (GM) crops presents a complex ethical landscape. We must carefully consider the potential benefits alongside the potential risks, engaging in a thoughtful and inclusive discussion that involves scientists, policymakers, farmers, and consumers alike. Failing to address these ethical concerns head-on could severely undermine the potential of genetic gastronomy to improve food security and nutrition.

A primary concern revolves around potential environmental impacts. The introduction of GM crops could disrupt existing ecosystems, potentially leading to unforeseen consequences for biodiversity. For example, the development of herbicide-resistant crops might inadvertently contribute to the evolution of herbicide-resistant weeds, necessitating even stronger, and potentially more harmful, herbicides. Therefore, rigorous environmental impact assessments and ongoing monitoring are crucial before widespread adoption of any GM crop. Furthermore, the potential for gene flow from GM crops to wild relatives needs careful evaluation to prevent unintended consequences for the natural environment.

Equally important are the socioeconomic implications of genetic gastronomy. The adoption of GM crops could exacerbate existing inequalities, potentially benefiting large corporations at the expense of small-scale farmers in developing countries. Intellectual property rights associated with GM crops also raise concerns about access and affordability. Ensuring equitable access to these technologies and preventing the monopolization of the food system is paramount for achieving food security for all. Transparency in research, development, and deployment is vital to building public trust and fostering equitable participation in this field.

Beyond the environmental and socioeconomic dimensions, ethical considerations also extend to the potential impact on human health. While rigorous testing is typically conducted before the release of GM crops, long-term effects on human health remain a source of debate and public concern. Continuing research and robust monitoring systems are crucial to identify and mitigate any potential health risks associated with the consumption of GM foods. Open communication and transparency are essential to address public anxieties and build confidence in the safety and efficacy of this technology.

Another vital ethical consideration relates to the potential for unintended consequences. Genetic modification is a powerful tool, and it is essential to proceed with caution and a deep understanding of the potential ramifications. The complexity of biological systems means that even with rigorous testing, unforeseen effects can emerge. Therefore, a precautionary approach is essential, prioritizing comprehensive risk assessments and ongoing monitoring of the impacts of GM crops on both the environment and human health. This proactive stance should be a cornerstone of responsible innovation in the field.

The principle of informed consent is critical in the context of genetic gastronomy. Consumers should have access to clear and accurate information about the genetic modification of foods, allowing them to make informed choices about their consumption. Mandatory labeling of GM products is a key aspect of fostering transparency and ensuring that consumers are empowered to exercise their right to choose. Education and public engagement initiatives are also important to foster a better understanding of the science behind genetic modification and to address common misconceptions.

Finally, the governance and regulation of genetic gastronomy must be robust and transparent. Strong regulatory frameworks are needed to ensure the safe and responsible development and deployment of GM crops, balancing the potential benefits with the need to protect human health and the environment. International cooperation is essential to harmonize regulations and avoid trade barriers while ensuring consistent high standards across different countries. Regular evaluation and adaptation of these regulatory frameworks are necessary to ensure they keep pace with advances in genetic modification technology. Only through careful ethical considerations and proactive governance can we harness the full potential of genetic gastronomy while minimizing potential risks.

2 Genetic Modification Techniques

Scientists employ several sophisticated techniques to alter the genetic makeup of crops. These range from the precise targeting of CRISPR-Cas9 to older, broader methods of gene transfer. Understanding these methods is crucial for appreciating the potential and limitations of genetic gastronomy.

CRISPR-Cas9, a revolutionary gene-editing tool, allows scientists to make highly specific changes to a plant's DNA. Think of it as a molecular scalpel, capable of removing, inserting, or altering individual genes with remarkable accuracy. This precision minimizes unintended consequences often associated with earlier techniques. The ease and efficiency of CRISPR-Cas9 have propelled its widespread adoption in agricultural research, accelerating the development of improved crop varieties.

Beyond CRISPR-Cas9, other gene editing methods exist, each

with its strengths and weaknesses. Zinc finger nucleases (ZFNs) and transcription activator-like effector nucleases (TALENs) are examples of earlier technologies that, while less precise than CRISPR-Cas9, still offer valuable tools for targeted gene modification. The choice of method depends on several factors, including the specific gene being targeted, the desired outcome, and available resources. Researchers carefully consider these factors to select the most appropriate approach for their research goals.

It's important to compare these modern techniques to traditional plant breeding methods. Traditional breeding relies on crossing plants with desirable traits, a process that can take years, even decades, to achieve the desired outcome. Furthermore, traditional breeding often leads to the transfer of many genes, some of which may be undesirable. In contrast, gene editing offers greater precision and speed, allowing for the targeted modification of specific genes while leaving the rest of the genome intact. This targeted approach significantly reduces the risk of introducing unintended traits, creating a more predictable outcome.

The safety and regulation of gene editing techniques are paramount. Rigorous safety assessments are conducted before any genetically modified crop is released into the market. These assessments examine potential risks to human health and the environment. International regulatory bodies such as the FDA and the EU's EFSA play a vital role in establishing safety standards and evaluating the risk profiles of genetically modified organisms. These regulatory processes ensure that the benefits of genetic modification are carefully weighed against potential risks, ensuring responsible development and deployment.

Understanding these genetic modification techniques provides a cornerstone of knowledge in genetic gastronomy. The future of food security hinges on the responsible and ethical application of these powerful tools, combining the precision of gene editing with the wisdom of sustainable agricultural practices. The development of nutritious and resilient crops holds immense promise, offering solutions to global hunger and malnutrition, while fostering a more sustainable and environmentally conscious food system. The careful and considered development and deployment of these techniques are essential for ensuring that they serve their intended purpose, enhancing food security and promoting a more sustainable future. The integration of these techniques into agricultural practices demands a balanced approach, one that carefully weighs scientific advancements against ethical considerations and societal needs. This balanced perspective is crucial for fostering public trust and ensuring that the benefits of genetic gastronomy are realized responsibly and equitably. A deep understanding of the technologies and their implications is necessary for participating in the critical discussions shaping the future of our food systems. This understanding isn't just limited to scientists; informed public discourse is vital for responsible innovation and integration of these technologies into our world. The potential benefits of these techniques are substantial, yet so is the responsibility of their implementation. Ultimately, the judicious application of these tools will not only enhance the nutritional value and resilience of crops but will also shape the future landscape of food production and consumption globally. This calls for a collaborative effort involving scientists, policymakers, and the public to ensure that these technologies are used sustainably and equitably to benefit all.

2.1 CRISPR-Cas9 Technology

CRISPR-Cas9, a revolutionary gene-editing tool, offers unprecedented precision in manipulating plant genomes. Unlike older methods, CRISPR allows scientists to target specific DNA sequences with remarkable accuracy, effectively acting as molecular scissors to cut and replace genes. This targeted approach minimizes unintended consequences, a significant advantage over earlier techniques. Its relative simplicity and affordability have made it accessible to numerous researchers worldwide, accelerating advancements in agricultural biotechnology.

The mechanism of CRISPR-Cas9 is elegantly simple, yet incredibly powerful. It utilizes a guide RNA molecule, designed to match a specific DNA sequence within the plant's genome, acting as a GPS system for the Cas9 enzyme. This enzyme, a protein with DNA-cutting capabilities, follows the guide RNA to the target location and makes a precise cut in the DNA double helix. The cell's natural repair mechanisms then kick in, either by directly rejoining the cut ends (leading to gene disruption) or by using a provided DNA template to replace the targeted sequence with a new gene.

Scientists are leveraging CRISPR-Cas9 to enhance various crop traits, contributing significantly to genetic gastronomy. For instance, improving nutrient content is a primary focus. Researchers are modifying genes responsible for the biosynthesis of essential vitamins and minerals, leading to crops richer in these vital components. This 'biofortification' approach aims to combat micronutrient deficiencies in populations reliant on limited dietary sources. Similarly, enhancing the nutritional profile of staple crops can significantly improve public health outcomes globally, reducing widespread malnutrition.

Another vital application involves bolstering crop resilience against environmental stresses. Pest resistance is a major focus, with CRISPR used to disable genes that make crops susceptible to insect damage. This reduces reliance on chemical pesticides, aligning with sustainable agricultural practices and minimizing environmental impact. Moreover, enhancing drought tolerance through gene editing allows crops to thrive in water-scarce regions, a crucial adaptation in a climate-changing world. This approach directly contributes to food security, especially in arid and semi-arid regions often vulnerable to food shortages. Imagine the impact of drought-resistant rice in areas currently facing severe water limitations.

The efficiency of CRISPR-Cas9 stands in stark contrast to traditional breeding methods. Traditional methods rely on random crossing of plants and selection over generations, a time-consuming process with unpredictable outcomes. CRISPR's precision allows for more directed modifications, shortening the breeding cycle and increasing the likelihood of obtaining the desired traits. Furthermore, it overcomes limitations inherent in traditional breeding, enabling the transfer of genes even between distantly related species, opening avenues for innovative crop improvement strategies not previously feasible.

However, CRISPR-Cas9 technology, like any powerful tool, comes with its share of challenges and ethical considerations. Off-target effects, where the Cas9 enzyme accidentally cuts at unintended sites in the genome, remain a concern. While advancements are minimizing these risks, rigorous testing and validation are essential before any genetically modified crop reaches the market. The ethical implications of gene editing in food production also require careful consideration, encompassing public perception, environmental safety, and equitable access to this transformative technology. Transparency and public discourse are pivotal for building trust and responsible innovation in this field.

Addressing these challenges requires a multidisciplinary approach involving scientists, regulators, policymakers, and the public. Open communication and rigorous safety assessments are crucial to ensure responsible development and deployment of CRISPR-Cas9 in agriculture. The potential benefits of this technology in improving food security and nutrition are substantial, but the responsible and ethical application of this powerful tool necessitates careful consideration of potential risks and societal implications. A proactive and collaborative approach will ensure CRISPR-Cas9 technology is harnessed for the betterment of humanity, while minimizing potential downsides. Ultimately, informed decision-making based on scientific evidence and ethical considerations will shape the future of genetic gastronomy. The future of sustainable food systems depends on it.

2.2 Gene Editing Methods

Scientists employ several methods to precisely alter a plant's genetic makeup. These techniques range from older, less precise methods to the highly specific tools available today. Understanding these methods is crucial for appreciating the potential and limitations of genetic gastronomy.

One of the earliest gene editing methods involved the use of Agrobacterium tumefaciens, a naturally occurring soil bacterium. This bacterium has the remarkable ability to transfer its DNA into plant cells, a process that scientists have harnessed to introduce desired genes into the plant genome. While effective, this method lacks the precision of newer techniques and can lead to unpredictable insertions of genetic material.

More advanced techniques offer greater control over the editing process. Zinc finger nucleases (ZFNs) and transcription activator-like effector nucleases (TALENs) are examples of engineered nucleases that can target specific DNA sequences for modification. These nucleases act like molecular scissors, cutting the DNA at a precise location, allowing for the insertion or deletion of genetic material. While offering increased precision compared to Agrobacterium-mediated transformation, they can be relatively costly and time-consuming to design and synthesize.

The revolutionary CRISPR-Cas9 system has emerged as a powerful and versatile gene editing tool. This technology utilizes a guide RNA molecule to direct the Cas9 enzyme, acting as molecular scissors, to a specific location within the genome. Its simplicity, affordability, and ease of use have made CRISPR-Cas9 the preferred method for many researchers, significantly accelerating gene editing research in various fields, including agriculture. CRISPR allows for precise gene knockouts, insertions, and even base pair substitutions, paving the way for highly targeted modifications in crops.

Each gene editing method presents unique advantages and disadvantages. Agrobacterium-mediated transformation is a relatively straightforward technique, but lacks precision. ZFNs and TALENs offer increased precision but are more complex and expensive. CRISPR-Cas9 technology provides an excellent balance between precision, efficiency, and cost-effectiveness, making it the leading method in contemporary genetic engineering. The choice of method depends on several factors, including the specific modification needed, the available resources, and the desired level of precision. Future developments will likely lead to even more refined and efficient gene editing tools.

Beyond these specific techniques, the broader context involves the integration of gene editing within existing agricultural practices. Researchers are constantly refining existing protocols to enhance efficiency and reduce costs. Optimization involves aspects such as optimizing the delivery system of genetic material into plant cells, refining the selection process for successfully modified plants, and carefully considering the impact of the modification on the overall plant physiology and yield.

Moreover, ethical considerations and regulatory frameworks are critical aspects of gene editing. Public perception and acceptance of genetically modified organisms (GMOs) play a significant role in the adoption of these technologies. The development of transparent, rigorous regulatory mechanisms is essential to ensure the safe and responsible implementation of gene editing in agriculture. The careful assessment of potential environmental risks, combined with clear communication to the public, is crucial for fostering trust and acceptance of these advancements.

The field of genetic gastronomy relies heavily on these gene editing methods to cultivate nutritious food. The choice of the most appropriate technique depends on the specific genetic modification goal, and the ongoing refinement of these technologies promises even more precise and efficient methods in the future. Successful implementation also requires careful consideration of broader contexts, such as ethical implications, regulatory aspects, and public perception. The continuous development and refinement of gene editing technologies will undoubtedly play a crucial role in addressing global food security challenges and enhancing the nutritional value of food crops. The ongoing research and development in this area promise a future where genetic gastronomy contributes significantly to a healthier and more sustainable food system for the world.

2.3 Traditional Breeding vs. GMOs

Traditional breeding methods, practiced for millennia, rely on the natural process of sexual reproduction to combine desirable traits from different plant varieties. Farmers select plants with advantageous characteristics—like higher yields or disease resistance—and cross-pollinate them, hoping the offspring inherit these traits. This process is time-consuming, often unpredictable, and can be hindered by the limited genetic diversity within a species. Furthermore, desirable traits may be linked to undesirable ones, making it challenging to isolate and selectively breed for only the positive attributes. The resulting plants are often genetically complex, making it difficult to understand the precise genetic basis of their improved traits, hence hampering future refinements. Consider, for instance, the painstaking efforts required to breed disease-resistant wheat varieties; countless generations of plants were screened and selected over many years before achieving a satisfactory outcome. This illustrates the significant limitations and length of time involved. This conventional approach, while effective in many instances, can be slow and cumbersome, especially when facing urgent agricultural challenges like climate change or rapidly evolving pests and diseases.

Genetic modification (GM), on the other hand, offers a more precise and efficient approach. Instead of relying on random recombination during sexual reproduction, GM technologies allow scientists to directly introduce specific genes into a plant's genome. This targeted approach offers greater control over the resulting traits, enabling the introduction of desirable characteristics without the unintended consequences often associated with traditional breeding. For example, a single gene responsible for pest resistance can be inserted into a crop, eliminating the need to breed for that trait through generations of crossing and selection. This targeted precision significantly accelerates the development of improved varieties. Moreover, GM technology allows scientists to introduce genes from completely unrelated species, expanding the genetic pool available for crop improvement beyond the limitations imposed by traditional breeding methods.

This opens up new possibilities for enhancing crop characteristics, especially valuable traits not found within the plant species itself.

It's crucial to recognize that neither method is inherently superior; their suitability depends heavily on the specific application and the desired outcome. Traditional breeding remains invaluable for certain applications, particularly when dealing with complex traits controlled by multiple genes or where a broad genetic base is needed to maintain adaptability. The simplicity and long history of established practice also makes it a readily available technology, requiring less specialized infrastructure and resources. It provides a readily accessible approach for many small-scale farmers across the globe. However, for specific, targeted improvements, especially those involving the introduction of entirely novel traits, GM technologies often prove faster, more efficient, and more precise. Think of enhancing nutritional content, like the insertion of genes responsible for increased vitamin A in rice, a task practically impossible using traditional breeding techniques within a reasonable timeframe.

The ongoing debate surrounding GM crops often overlooks the fundamental differences between these two approaches. The public perception of GM is heavily influenced by factors like the speed of innovation and perception of risk, often without a complete understanding of the established processes of traditional breeding. It's easy to overlook the numerous challenges associated with traditional techniques, which often involve using chemicals and extensive labor. Therefore, a holistic understanding of both methods, recognizing their strengths and limitations, is essential for informed decision-making regarding food production and agricultural practices. Ultimately, a balanced approach, using both traditional breeding and GM techniques where appropriate, holds the most promise for developing sustainable and resilient agricultural systems to meet global food security needs. The most effective path likely involves a combination of both, leveraging the advantages of each approach depending on the specific context and desired outcome. Such a blend allows for a flexible and innovative approach to developing improved crop varieties that simultaneously address food security concerns and promote ecological sustainability.

3 Enhancing Nutritional Value

Genetically modified (GM) crops offer a powerful tool to enhance the nutritional value of our food supply, addressing widespread micronutrient deficiencies and improving overall public health. Biofortification, a key strategy, focuses on increasing the levels of essential vitamins and minerals directly within the plant itself, unlike fortification which adds nutrients externally after harvest. This inherent increase offers a more sustainable and accessible solution, particularly in regions with limited access to processed foods and supplements.

Consider the example of Golden Rice, genetically engineered to produce beta-carotene, a precursor to vitamin A. Millions worldwide suffer from vitamin A deficiency, leading to blindness and increased susceptibility to infections. Golden Rice directly addresses this problem by providing a readily available source of this crucial vitamin through a staple food. The impact of such innovations cannot be overstated in improving the health and well-being of vulnerable populations.

However, biofortification isn't simply about adding vitamins; it encompasses a broader approach to improving nutritional profiles. Scientists are actively working on enhancing the levels of essential minerals like iron and zinc in staple crops, aiming to combat deficiencies that impair cognitive development and immune function. These efforts require a multidisciplinary approach, combining genetic engineering expertise with agricultural science, nutritional research, and public health strategies. Successful implementation depends on factors far beyond the laboratory—consumer acceptance, efficient distribution networks, and supportive government policies all play crucial roles.

Furthermore, the potential of GM crops extends beyond single nutrient enhancements. Researchers are exploring ways to improve the bioavailability of existing nutrients – ensuring the body can efficiently absorb and utilize them. This aspect is often overlooked, but it is critical to maximizing the nutritional impact of any modification. For instance, the genetic makeup of a plant can influence the form of iron it contains, impacting how readily the human body absorbs it. Modifying genes involved in iron metabolism can therefore significantly increase the nutritional benefits, rendering the iron more bioavailable. This research necessitates collaboration across different scientific fields, integrating plant biology, human nutrition, and even molecular biology to gain a more comprehensive understanding.

The benefits of biofortification extend beyond direct health improvements; they also contribute to sustainable food systems. By increasing the nutritional density of crops, we reduce the need for extensive supplementation programs. These programs often face logistical challenges, require significant resources, and can be unsustainable in the long term. Biofortification offers a more integrated and self-sufficient solution, embedding nutritional improvements directly within the food production system. This has far-reaching consequences, promoting food security and reducing the reliance on external interventions.

In the future, we can expect further advancements in biofortification. Advanced gene-editing technologies like CRISPR-Cas9 allow for greater precision and efficiency in modifying crop genomes. This enables the simultaneous improvement of multiple nutritional traits, leading to a more holistic enhancement of food quality. For instance, we could potentially engineer crops with increased levels of both iron and zinc, simultaneously addressing two major micronutrient deficiencies. Moreover, emerging technologies promise more refined control over metabolic pathways, optimizing nutrient production and bioavailability further. This makes the creation of truly nutritionally enhanced crops a realistic and achievable goal.

Creating nutritionally enhanced crops through genetic modification is not without challenges. Ethical considerations, public perception, and regulatory hurdles continue to be factors influencing the adoption and implementation of these technologies. Open communication, public education, and transparent regulatory processes are all essential to addressing these concerns and fostering widespread acceptance of biofortification. Furthermore, rigorous safety assessments must be conducted to ensure that GM crops are safe for human consumption and do not have unintended environmental consequences. Addressing these multifaceted aspects is crucial for successfully leveraging the full potential of GM crops to enhance nutritional value and contribute to global health and food security. The advancements in technology combined with ongoing research in human nutritional needs offer a promising pathway to a healthier and more secure future. It is through a collaborative effort, integrating scientific innovation with careful ethical considerations and sound public policy that we can fully realize the potential of biofortification and genetic modification to feed and nourish the world. The future of food security is intrinsically linked to the acceptance and implementation of these groundbreaking technologies. Therefore, understanding their potential benefits and addressing any concerns associated with them is of paramount importance in creating a future where everyone can access sufficient, safe, and nutritious food.

3.1 Fortified Crops

Golden rice, enriched with beta-carotene, a precursor to Vitamin A, stands as a powerful testament to the potential of fortified crops. This crucial vitamin, vital for vision and immune function, is deficient in the diets of millions globally, predominantly in developing nations. Golden rice offers a direct, targeted solution, enhancing the nutritional value of a staple crop directly within the grain itself, bypassing reliance on external supplementation strategies, often hampered by access and affordability. The technology behind its creation showcases the precision of modern genetic engineering techniques, allowing for the introduction of specific genes without disrupting the plant's overall biology. Further research continues to refine golden rice, increasing beta-carotene levels and tailoring it to specific regional conditions, for maximum impact and widespread adoption.

Beyond beta-carotene, fortified crops encompass a broader spectrum of essential nutrients. Iron-fortified beans and lentils, for instance, directly address iron deficiency, another significant global health issue. Iron is crucial for oxygen transport and energy production, and deficiency leads to anemia, impacting cognitive function and physical development. The fortification process involves introducing genes that enhance the plant's ability to absorb and accumulate iron during its growth cycle, resulting in a product naturally richer in this essential mineral. Unlike supplements, fortified foods are seamlessly integrated into existing diets, circumventing compliance issues and cultural barriers, ensuring a more effective and sustainable impact on public health.

The development of fortified crops is not without its complexities. Careful consideration is needed to ensure the nutritional enhancement does not compromise other aspects of the crop's quality, including taste, texture, and yield. Rigorous testing and field trials are vital, evaluating the fortified crop's performance under diverse environmental conditions and assessing its impact on the surrounding ecosystem. Moreover, engaging local communities in the development and implementation of these crops is paramount, ensuring cultural sensitivity and facilitating widespread acceptance and adoption. Effective communication of the benefits, addressing any potential concerns, are essential for successful integration into food systems.

Furthermore, the long-term sustainability of fortified crops hinges on their economic viability for farmers. High yields and cost-effective production are crucial for widespread cultivation and distribution. This necessitates collaboration between researchers, agricultural extension services, and policymakers to develop appropriate support systems, including training, access to quality seeds, and market incentives. Creating supportive agricultural policies is essential for making fortified crops profitable and accessible to farmers. Only with such comprehensive strategies can these crops realize their full potential in improving global nutrition.

Genetically modified crops are not simply about boosting nutritional content; they also offer a crucial strategy for enhancing food security in the face of growing environmental challenges. For example, drought-tolerant varieties of staple crops are essential in regions facing water scarcity. Genetically engineered crops with enhanced drought resistance can continue to produce viable yields even under limited water availability, contributing to more reliable and consistent food production in vulnerable areas. This is particularly critical for regions facing increasing aridity due to climate change, ensuring that food production remains resilient in the face of environmental stresses. Reducing reliance on irrigation is a boon for both environmental sustainability and economic efficiency.

The creation of fortified crops represents a powerful synergy between technological advancement and the pursuit of global food security. It showcases the capacity of genetic engineering to directly address critical nutritional deficiencies and enhance the resilience of food systems. However, successful implementation demands a multifaceted approach, encompassing scientific research, effective policy frameworks, farmer empowerment, and widespread public acceptance. The ultimate goal is to improve the nutritional status of vulnerable populations while establishing sustainable, resilient food production systems for generations to come. This multifaceted approach, focusing on research, regulation, and education, will be key to unlocking the full potential of fortified crops. The ongoing research continues to refine and expand this technology, creating crops fortified with a wider range of essential nutrients and tailored to specific regional needs.

3.2 Biofortification Strategies

Biofortification isn't just about adding vitamins to processed foods; it's about fundamentally altering the nutritional profile of crops at their source. Imagine a world where staple crops like rice and maize naturally contain significantly higher levels of essential vitamins and minerals, improving the diets of billions. This is the power of biofortification, a crucial strategy in genetic gastronomy. We aren't talking about superficial enhancements; we're discussing genetically modifying plants to inherently produce more nutrients. This is a game-changer for global nutrition. This approach offers a far more sustainable and equitable solution than simply fortifying processed foods. Fortified foods often rely on synthetic vitamins and minerals, may not be accessible to all populations, and can be prone to nutrient loss during processing and storage. Biofortification, however, integrates the desired nutrients directly into the plant's genetic makeup, ensuring the nutrients are present throughout the entire plant lifecycle from seed to consumption. This means more consistent nutrient delivery, regardless of environmental factors or processing methods. The benefits extend beyond simple nutrient enrichment; it contributes to enhanced food security and public health.

Consider the example of Golden Rice, engineered to produce beta-carotene, a precursor to vitamin A. Millions suffer from vitamin A deficiency worldwide, leading to blindness and increased mortality, particularly among children. Golden Rice offers a potential solution, providing a readily accessible source of vitamin A directly through a staple food. However, it's critical to understand that biofortification isn't a silver bullet. The effectiveness of biofortified crops depends on factors such as consumer acceptance, cultivation practices, and effective distribution networks. It requires a holistic approach, integrating agricultural research, public health initiatives, and socio-economic considerations.

Successful biofortification strategies involve precise genetic engineering techniques, often employing gene editing tools like CRISPR-Cas9. Scientists carefully identify genes responsible for nutrient biosynthesis and modify them to enhance nutrient production. This is not a process of simply adding genes; it's about fine-tuning existing metabolic pathways to increase the efficiency of nutrient synthesis within the plant. Researchers meticulously test the modified crops for safety and efficacy, ensuring they meet stringent regulatory standards before release for commercial cultivation. The rigorous scientific process emphasizes the safety and reliability of the biofortification technology and dispels any unfounded concerns.

Beyond vitamin A, biofortification targets other vital nutrients. Scientists are working on enhancing the iron content of staple crops, combatting iron deficiency anemia, a widespread problem, especially among women and children. Zinc deficiency is another global health challenge, and biofortification strategies are being developed to enhance the zinc content of rice, beans, and other important crops. Biofortification aims to address multiple nutrient deficiencies simultaneously, providing a holistic approach to improving nutritional status and overall health. The long-term goal is to fortify multiple nutrients in widely consumed crops, creating a substantial impact on global health and reducing reliance on costly and potentially unsustainable fortification programs.

The development of biofortified crops requires careful consideration of the specific nutritional needs of the target population. It's crucial to understand local dietary habits, food consumption patterns, and the prevalence of specific micronutrient deficiencies. A successful biofortification program must be tailored to the needs of the community it intends to serve. Furthermore, biofortification strategies should be environmentally sustainable, minimizing the impact on biodiversity and promoting environmentally friendly farming practices. The aim is not just to improve nutrition but to create a more resilient and sustainable food system. The effective and responsible implementation of biofortification necessitates a deep understanding of both the biological and socio-economic dimensions of the problem, to ensure that the solutions are truly sustainable and effective in improving human health and well-being. This requires close collaboration among scientists, policymakers, farmers, and consumers.

Furthermore, the acceptance and adoption of biofortified crops among farmers and consumers are pivotal for the success of these initiatives. Clear communication regarding the benefits of biofortified crops and addressing any safety concerns are vital. Strategies to encourage farmers' adoption should include access to high-quality seeds, appropriate training, and supportive policies that incentivize the cultivation and distribution of biofortified crops. Likewise, educational campaigns aimed at raising consumer awareness and promoting the acceptance of biofortified crops are essential to enhance their utilization and maximize the intended public health benefits. The ultimate success depends upon a collaborative effort to ensure widespread adoption and effective integration into existing food systems. This inclusive approach, encompassing diverse stakeholders, is crucial to achieving the full potential of biofortification and achieving significant improvements in global nutrition.

4 Improving Crop Resilience

Developing pest-resistant crops is crucial for sustainable agriculture. Traditional methods often involve broad-spectrum pesticides, harming beneficial insects and potentially polluting water sources. Genetically modified (GM) crops, however, can incorporate genes from other organisms that confer natural resistance to specific pests. This targeted approach reduces the need for chemical pesticides, leading to a healthier environment and more sustainable farming practices. For instance, Bt crops, engineered to produce insecticidal proteins from Bacillus thuringiensis, effectively control pests like corn borers and bollworms, minimizing pesticide use and maximizing crop yield. The benefits extend beyond environmental protection; reduced pesticide costs also improve the economic viability of farming for growers, particularly smallholder farmers in developing countries.

Moreover, the reduced reliance on pesticides minimizes the risk of pesticide resistance developing in pest populations, ensuring long-term effectiveness of pest control strategies. Finally, the increased yields obtained through pest resistance contributes to greater food security, particularly in regions vulnerable to pest infestations. The development of pest-resistant GM crops represents a significant advancement in sustainable agriculture, balancing high yields with environmental protection and economic benefits.

Herbicide tolerance in GM crops offers another avenue for improving crop resilience. Conventional weed control methods rely heavily on herbicides, some of which can have detrimental effects on the environment and human health. GM crops engineered for herbicide tolerance allow farmers to utilize herbicides more efficiently. Specific herbicides can be applied to control weeds without harming the crop itself, reducing the overall amount of herbicide needed. This targeted approach minimizes herbicide drift, protects non-target organisms, and lessens the environmental burden associated with large-scale herbicide application. Furthermore, herbicide-tolerant crops can simplify weed management, reducing labor costs and potentially increasing crop yields by minimizing weed competition for resources. However, concerns remain regarding the potential for herbicide-resistant weeds to emerge, highlighting the need for integrated pest management strategies that combine herbicide-tolerant crops with other weed control methods. The careful consideration of herbicide application rates and the adoption of diversified farming practices are essential to mitigate the potential risks associated with herbicide-tolerant GM crops and ensure their sustainable use.

Drought tolerance is becoming increasingly important as climate change exacerbates water scarcity in many agricultural regions. GM technology offers the potential to develop crops that can withstand periods of drought more effectively. Scientists are working on enhancing the efficiency of water use in plants through genetic modification, allowing them to survive and produce yields even under water-stressed conditions. This involves altering genes that regulate water uptake, transportation, and retention, potentially improving plant physiology to withstand drier conditions. Furthermore, GM crops engineered for drought tolerance could significantly enhance agricultural productivity in arid and semi-arid regions, increasing food security in areas particularly vulnerable to climate change impacts. However, the development of drought-tolerant crops is a complex process, often requiring the integration of multiple genes and careful consideration of interactions with other environmental factors. Success depends on thorough field testing and careful adaptation to specific environmental conditions to ensure that the benefits are realized while risks are minimized.

Disease resistance in crops is another area where GM technology offers significant potential. Many diseases can devastate crops, reducing yields and potentially causing widespread famine. Through genetic engineering, it is possible to incorporate genes that provide resistance to specific pathogens, thereby protecting crops from disease and enhancing their productivity. This can reduce reliance on chemical fungicides, contributing to a more environmentally friendly approach to agriculture. Moreover, enhanced disease resistance can reduce crop losses, improving food security and farmer livelihoods. However, the development of disease-resistant crops requires careful consideration of the potential for pathogens to evolve and overcome the introduced resistance. Thus, comprehensive disease management strategies that combine genetic resistance with other disease control measures are essential for long-term success. The integration of genetic engineering with other sustainable agricultural practices remains a crucial aspect of ensuring resilience against crop diseases.

Adapting crops to climate change is essential for ensuring food security in a changing world. GM technology can play a significant role in this adaptation, by engineering crops that can withstand a wider range of environmental conditions, including increased temperatures, altered rainfall patterns, and elevated carbon dioxide levels. This could involve improving heat tolerance, enhancing drought resistance, or altering photosynthetic pathways to optimize resource use under changing climate conditions. The integration of different stress tolerance traits into a single crop offers a powerful strategy for creating more resilient varieties capable of surviving and yielding well in diverse climates. However, the development of climate-resilient crops requires careful consideration of the complexities of climate change, including unpredictable and variable weather patterns. The long-term efficacy of GM crops in mitigating climate change effects also needs to be carefully assessed, considering potential interactions with other environmental factors. The process involves multidisciplinary collaboration among geneticists, agronomists, and climate scientists.

4.1 Pest Resistance

Developing pest-resistant crops is crucial for sustainable agriculture. Traditional methods often involve the use of broad-spectrum pesticides, which can harm beneficial insects and contaminate the environment. Genetically modified (GM) crops offer a more targeted approach. Scientists can insert genes from other organisms, often bacteria, into the plant's DNA, conferring resistance to specific pests. This reduces the need for chemical pesticides, minimizing environmental damage and promoting biodiversity. The Bacillus thuringiensis (Bt) gene, for instance, produces a protein toxic to certain insects, effectively protecting the plant.

The development of pest-resistant GM crops involves a complex process. Initially, scientists identify the target pest and the gene responsible for its vulnerability. They then isolate and insert this gene into the plant's genome using various gene editing techniques, such as CRISPR-Cas9, which allow for precise and efficient gene modification. Rigorous testing is conducted to evaluate the effectiveness of the gene in providing pest resistance, ensuring the absence of unintended effects on the plant's overall health, and assessing its impact on non-target organisms.

One key benefit of pest-resistant GM crops is the reduction in pesticide use. This translates to lower costs for farmers, reduced environmental pollution, and a safer working environment. Furthermore, minimizing pesticide application contributes to the preservation of beneficial insect populations, enhancing ecosystem health. The decreased reliance on chemical pesticides also safeguards human health by reducing exposure to potentially harmful substances found in such pesticides. The overall impact on the environment, from soil and water quality to biodiversity, is positively affected by the reduction in chemical applications.

However, the development and implementation of pest-resistant GM crops also present challenges. One major concern is the potential for pest resistance to evolve over time. Continuous use of a single pest-resistance gene could lead to the emergence of pest populations that develop resistance, negating the intended effect. This risk necessitates the development of integrated pest management strategies, combining GM crops with other methods to prevent the development of resistance. Researchers are exploring diverse strategies to mitigate this issue, including the development of crops with multiple resistance genes or the use of different pest control methods in rotation.

Another challenge revolves around public perception and acceptance. Concerns over potential ecological impacts and long-term health effects have led to public resistance against GM crops in some regions. Clear and transparent communication of the scientific evidence supporting the safety and benefits of pest-resistant GM crops is essential to alleviate public concerns. It's crucial to address the ethical, environmental, and economic implications thoroughly and openly in the dialogue. Education and engagement with the public are crucial steps to overcoming these obstacles. Increased transparency and scientific communication can pave the way for wider acceptance and the successful adoption of these innovative technologies.

The development of pest-resistant GM crops represents a significant advancement in sustainable agriculture. While challenges remain, the potential benefits, including reduced pesticide use, improved crop yields, and enhanced environmental protection, make it a promising tool for ensuring global food security. Addressing the challenges necessitates a multi-faceted approach incorporating ongoing research, rigorous testing, effective communication, and responsible stewardship of this technology. It is only through addressing these challenges responsibly that this technology will flourish and contribute to feeding the world more sustainably. The future holds the promise of continued innovation in this field, leading to even more effective and sustainable pest control solutions. The continued development and responsible use of pest-resistant GM crops holds immense potential for positive impacts on agriculture and the environment.

4.2 Herbicide Tolerance

Herbicide tolerance in genetically modified (GM) crops represents a significant advancement in agricultural technology. This characteristic, achieved through genetic engineering, allows farmers to effectively control weeds without harming the target crop, ultimately leading to higher yields and reduced reliance on potentially harmful herbicides. The process involves introducing genes that confer resistance to specific herbicides, enabling the crop to withstand application of these chemicals while effectively eliminating competing weeds. This strategy is particularly valuable in managing weed infestations, which significantly impact crop productivity and quality.

Developing herbicide-tolerant crops requires a meticulous understanding of plant physiology and genetics. Scientists identify genes that confer herbicide resistance, often sourced from naturally occurring herbicide-resistant plants or bacteria. These genes are then incorporated into the genome of the target crop using various genetic engineering techniques, such as Agrobacterium-mediated transformation or gene editing tools like CRISPR-Cas9. Rigorous testing is crucial to ensure the introduced genes function correctly and don't negatively impact other plant characteristics or pose ecological risks. The efficiency and precision of these techniques are constantly improving, leading to more effective and targeted herbicide tolerance. The benefits of herbicide tolerance extend beyond increased crop yields. Reduced herbicide use contributes to a more sustainable agricultural system by minimizing environmental impact. Conventional herbicide application can lead to soil erosion, water contamination, and harm to beneficial insects and other organisms. Herbicide-tolerant crops allow for reduced herbicide use, mitigating these negative effects and promoting biodiversity. Furthermore, the decreased reliance on herbicides reduces the associated costs for farmers, improving their economic viability.

However, the widespread adoption of herbicide-tolerant crops also presents challenges. The development of herbicide-resistant weeds, a phenomenon known as "superweeds," is a major concern. Prolonged exposure to a particular herbicide can select for weeds with inherent or acquired resistance, rendering the herbicide ineffective. To counteract this, integrated pest management strategies combining various weed control methods, including crop rotation and mechanical weeding, are crucial. Furthermore, careful consideration must be given to the potential for gene flow from herbicide-tolerant crops to wild relatives, potentially leading to the emergence of herbicide-resistant weeds in natural ecosystems.

Effective management of herbicide-tolerant crops requires a comprehensive approach that integrates various aspects of agricultural science. Farmers need thorough training on the proper use and application of herbicides, emphasizing best practices to minimize risks and maximize effectiveness. Additionally, the development and implementation of monitoring programs to track the emergence and spread of herbicide-resistant weeds are essential for early detection and effective management strategies. Ongoing research focuses on developing alternative herbicides with different modes of action, reducing the reliance on a single herbicide and minimizing the selection pressure for resistance. This multifaceted approach ensures the long-term sustainability of herbicide tolerance as a key agricultural technology.

The responsible use of herbicide-tolerant crops plays a vital role in ensuring food security and environmental sustainability. By integrating herbicide tolerance with other sustainable agricultural practices such as crop rotation, cover cropping, and integrated pest management, farmers can minimize the ecological impact while maximizing crop production. Careful consideration of potential risks, such as the development of herbicide-resistant weeds and potential gene flow, is necessary for mitigating negative consequences. Continued research and technological advancements will refine these strategies and further optimize the benefits of herbicide tolerance in supporting a robust and sustainable agricultural system that meets the demands of a growing global population. The future of agriculture increasingly relies on innovative and sustainable approaches, with herbicide-tolerant crops playing a significant, albeit carefully managed, role in global food security. Responsible development and implementation of these technologies are crucial for achieving long-term agricultural sustainability.

4.3 Drought Tolerance

Developing drought-tolerant crops is crucial for ensuring global food security, especially in arid and semi-arid regions. These regions are projected to experience more frequent and intense droughts due to climate change, severely impacting agricultural yields and livelihoods. Genetic modification offers a powerful tool to enhance a plant's ability to withstand water scarcity. This involves manipulating genes that regulate water uptake, retention, and stress response mechanisms.

One promising strategy involves modifying genes related to stomatal control. Stomata are tiny pores on plant leaves that regulate gas exchange and water transpiration. By engineering plants with improved stomatal regulation, we can minimize water loss while maintaining sufficient photosynthesis. This could involve altering the expression of genes encoding key proteins in the stomatal apparatus, such as those responsible for stomatal opening and closure. Research in this area has shown that genetically modified plants with modified stomatal conductance can significantly improve drought tolerance.

Another avenue involves enhancing root systems. Drought-resistant plants often possess extensive and deeply penetrating root systems, allowing them to access water reserves deep within the soil. Genetic engineering can be used to modify genes involved in root development, leading to improved root architecture, increased root biomass, and enhanced water absorption. For example, manipulating genes responsible for root branching and root hair development could lead to a more efficient root system, allowing the plant to extract more water from the soil during periods of drought. These modifications can be coupled with other genetic changes to improve overall drought stress tolerance.

Beyond root systems and stomatal control, we can improve stress tolerance through the manipulation of genes involved in osmoprotection. Osmoprotectants are molecules that help plants maintain osmotic balance under water stress conditions. Genetic engineering can enhance the production of these osmoprotectants, thus improving the plant's ability to withstand drought stress. This approach could involve introducing genes encoding enzymes responsible for the synthesis of osmoprotectants or overexpressing existing genes to enhance their production.

Furthermore, the use of genetic engineering to enhance the plant's ability to regulate abscisic acid (ABA) is also pivotal. ABA is a plant hormone that plays a crucial role in drought stress response. By manipulating genes involved in ABA biosynthesis and signaling pathways, we can enhance the plant's ability to activate drought stress tolerance mechanisms efficiently. This may involve increasing ABA production during drought conditions, leading to more effective stress response.

The development of drought-tolerant crops isn't simply a matter of enhancing individual traits in isolation; it requires an integrated approach. Multiple gene modifications are often necessary to achieve substantial improvements in drought tolerance. This approach, often termed "gene stacking," involves combining multiple beneficial traits into a single plant variety. This approach might involve combining improved stomatal control, enhanced root systems, and increased osmoprotectant production. The resulting plant exhibits significantly higher drought resistance than those with single modifications.

It's also crucial to consider the broader environmental implications of these modifications. The goal is not simply to create drought-tolerant crops, but also to ensure that these crops do not have unintended negative consequences for the environment. This necessitates thorough risk assessment and careful field testing before widespread adoption. It's important to conduct thorough research and development to ensure that new traits do not negatively impact biodiversity or ecological balance. A rigorous and transparent approach to risk assessment is paramount to safeguard the environment.

Finally, successful implementation of drought-tolerant GM crops depends on various factors, including public perception and regulatory frameworks. Effective communication about the benefits and safety of GM technology is vital to garner public acceptance and ensure the successful adoption of these crops by farmers. Transparency and open dialogue about the environmental and economic impacts of GMOs are important for effective adoption. Additionally, supportive regulatory frameworks are essential to encourage further research, development, and deployment of drought-tolerant GM crops. A combination of robust regulatory approval processes and effective communication about their benefits and safety is essential for successful integration.

4.4 Disease Resistance

Disease resistance in crops is paramount for ensuring food security and minimizing economic losses. Farmers worldwide face significant challenges from various pathogens, including fungi, bacteria, and viruses, leading to substantial yield reductions and impacting livelihoods. Genetic modification offers a powerful tool to combat these threats, moving beyond traditional methods which often prove insufficient against rapidly evolving pathogens. The development of disease-resistant varieties can significantly reduce the reliance on chemical pesticides, contributing to a more sustainable agricultural system. This is crucial not just for environmental reasons, but also because many effective pesticides are becoming less effective due to the development of resistance in pest populations.

Developing disease-resistant crops involves a multi-faceted approach. Scientists may identify specific genes within a plant's genome responsible for its resistance against a particular disease. These genes can then be isolated and introduced into other crop varieties lacking this inherent resistance, effectively bestowing disease resistance upon them. CRISPR-Cas9 technology plays a significant role in this precision gene editing; allowing for the targeted introduction or modification of specific genes associated with disease resistance, which enables tailoring the genetic changes to minimize unintended consequences and maximize efficacy. For example, scientists have successfully integrated genes from wild relatives of cultivated crops, known for their natural disease resistance, enhancing the resistance of popular and commercially successful varieties.

However, the path to achieving durable disease resistance is not always straightforward. Pathogens can evolve rapidly, adapting to overcome the introduced resistance mechanisms. This necessitates a continuous evaluation of the effectiveness of the modified crops and development of new strategies to maintain resistance over time. In fact, constant monitoring and refinement of the genetic modifications, coupled with integrated pest management strategies that incorporate other disease control methods, are paramount for long-term success. For example, the implementation of diverse cropping systems can reduce pest pressure and slow the spread of diseases, complementing the resistance traits introduced through genetic modification. The goal isn't merely to create disease-resistant plants, but also a system that minimizes disease risk and maximizes the long-term productivity of the plants.

Furthermore, the public perception of genetically modified organisms (GMOs) significantly influences the adoption of disease-resistant crops. Concerns about the potential environmental impacts and human health effects need to be addressed through transparent communication and robust risk assessments. It is crucial to emphasize the benefits of reducing pesticide use and enhancing environmental sustainability, thereby showing the positive contribution of these crops to broader ecological health. Open dialogue and increased public understanding are crucial for facilitating the widespread acceptance and successful integration of these technological advancements into agricultural practices. The focus should be on explaining the science behind these technologies, highlighting the potential environmental and social benefits, and fostering public trust through transparency and scientific rigor.

Beyond the technical aspects, economic factors play a critical role in the successful implementation of disease-resistant crops. The cost of developing and commercializing genetically modified varieties can be substantial, requiring significant investments in research, development, and regulatory approvals. Therefore, policies that support innovation and ensure fair access to these technologies are vital for maximizing their impact on food security, especially in developing countries where the burden of crop diseases is often highest. Furthermore, supportive policies regarding farmer access and affordable seed production ensure the equitable distribution of these benefits among farmers, enhancing their resilience to crop failure and improving overall food security at a community level. A comprehensive approach that considers scientific advancements, economic realities, and public acceptance is crucial for harnessing the full potential of genetic modification to improve disease resistance in crops.

4.5 Climate Change Adaptation

Adapting crops to withstand the escalating effects of climate change is crucial for ensuring global food security. Genetically modified (GM) crops offer a powerful tool in this fight, allowing us to engineer resilience into our food sources. We're not just talking about survival; we're talking about thriving in the face of unpredictable weather patterns and increasingly hostile growing conditions. This isn't about replacing traditional farming; it's about augmenting its capabilities with precision tools tailored to specific environmental challenges.

The development of drought-tolerant varieties is paramount. Imagine fields of corn, typically thirsty plants, surviving extended periods without significant rainfall. This isn't science fiction; researchers are actively incorporating genes from drought-resistant species into major crops, drastically reducing water needs and enabling cultivation in previously arid regions. This increased resilience translates directly to reduced reliance on irrigation, conserving precious water resources and lowering agricultural costs. Furthermore, this means access to vital food sources in regions particularly vulnerable to water scarcity, potentially mitigating food shortages and preventing conflict over diminishing resources.

Similarly, heat tolerance is being engineered into crops, enabling them to withstand extreme temperatures without significant yield reduction. Heat stress affects plant growth and reproductive capabilities, leading to lower crop yields and impacting food production. By modifying genes responsible for heat-shock proteins—proteins that help protect cells under stress—we can develop crops that can endure increasingly frequent and intense heat waves. This targeted genetic modification, unlike broad-spectrum pesticide use, introduces minimal disruption to the ecosystem and focuses specifically on enhancing the plant's inherent ability to survive.

Pest and disease resistance is another key area. Climate change alters the distribution and behavior of pests and pathogens, increasing the incidence of crop diseases and necessitating increased pesticide use. GM crops can be engineered to resist specific pests and diseases, reducing or eliminating the need for chemical interventions. This cuts down on environmental damage from pesticide runoff and improves the overall health of agricultural ecosystems. Moreover, this translates to lower production costs for farmers, helping to ensure the affordability of food products even in times of heightened stress.

Beyond these direct adaptations, there are more subtle, yet equally important, genetic enhancements we're exploring. We can modify crops to optimize photosynthesis efficiency, enabling them to maximize the use of available sunlight even under less-than-ideal conditions. This approach leads to increased yield without relying on additional resources. This translates to higher production rates, greater food availability, and increased economic sustainability for farmers. These are not isolated advancements; the cumulative effect is transformative, creating a resilient and robust agricultural system capable of meeting global demands.

The successful implementation of climate-resilient GM crops requires careful consideration. Effective regulatory frameworks must balance innovation with safety, fostering technological progress while mitigating potential risks. Open communication with the public is vital to building trust and addressing concerns regarding the safety and ethical implications of genetic modification. This involves transparent research, engaging dialogue with stakeholders, and ensuring that the benefits of this technology reach those most in need. This multifaceted approach, encompassing science, regulation, and public engagement, is paramount to harnessing the full potential of GM technology in mitigating the impacts of climate change and ensuring sustainable food production for future generations. The stakes are high, but the potential rewards—a more secure and sustainable food system—are worth the effort. The future of food security depends, in part, on our ability to cultivate crops that thrive in a changing world.

5 Sustainable Farming Practices

Genetically modified (GM) crops offer a powerful tool for minimizing environmental impact in agriculture. Pest-resistant varieties, for instance, significantly reduce the need for synthetic pesticides, lessening the associated risks to biodiversity and human health. This decrease in pesticide application translates directly to cleaner water sources and healthier soil ecosystems, benefiting not only the immediate agricultural environment but also adjacent natural habitats. Furthermore, the reduced reliance on pesticides contributes to a decrease in greenhouse gas emissions associated with their production and transportation. The overall impact is a substantial move toward a more ecologically responsible agricultural system.

Precision breeding techniques, including gene editing, enable the development of crops better adapted to specific environmental conditions. Drought-tolerant varieties, for example, drastically reduce water consumption, a critical factor in arid and semi-arid regions increasingly facing water scarcity. This decreased water usage not only conserves precious resources but also minimizes the energy required for irrigation, thereby further reducing the carbon footprint of food production. These advancements offer substantial improvements in water management efficiency. Such strategies are paramount for ensuring food security and agricultural sustainability in the face of climate change.

Beyond pest and drought resistance, genetically modified crops can play a key role in combating soil degradation. Certain GM crops exhibit improved nutrient uptake efficiency, reducing the need for synthetic fertilizers which contribute to soil erosion and nutrient depletion. The reduction in fertilizer use translates to less runoff contaminating water systems, improving water quality and reducing associated health risks. This shift towards more efficient nutrient use also benefits soil health, leading to more productive and sustainable farming practices over the long term, reinforcing the positive effects on the ecosystem. It's important to note that soil health is fundamental to food production and overall environmental sustainability.

Sustainable farming necessitates efficient resource management, and GM crops can contribute significantly. For example, crops engineered for improved nitrogen fixation, a crucial process in plant growth, can significantly decrease the reliance on nitrogen-based fertilizers. This reduction in fertilizer dependence not only lessens the environmental impact but also can lower production costs for farmers, making farming more economically sustainable. The potential economic benefits also extend to the consumers, possibly lowering food prices. Farmers can achieve higher yields with less input, promoting both environmental sustainability and economic viability.

The long-term effects of reduced pesticide and fertilizer usage on biodiversity are significant. By minimizing the use of chemical inputs, farmers contribute directly to the preservation of beneficial insects, pollinators, and other crucial components of healthy ecosystems. The presence of diverse plant and animal species, in turn, promotes a more resilient and robust agricultural landscape, better equipped to withstand environmental challenges. It is this integrated ecosystem approach that underpins sustainable farming practices and ensures the long-term health of our planet.

Integrating genetic modification into sustainable farming practices also involves considering the entire production cycle. This includes optimizing planting and harvesting techniques to minimize soil compaction and erosion, which are often worsened by heavy machinery. Furthermore, precision agriculture tools, often informed by data from genetically modified crops, allow for more efficient use of resources, reducing waste and optimizing inputs throughout the entire farming process. This holistic perspective ensures the sustainability of farming practices beyond just the use of GM crops.

The development and implementation of GM crops are not without challenges. Thorough research and comprehensive risk assessments are crucial to ensure the safety of these technologies and to mitigate any potential unintended consequences. Transparency and open communication regarding the application of GM technologies are essential to building public trust and confidence. Collaborative approaches involving scientists, policymakers, and farmers are paramount to successful implementation and sustainable integration into agricultural systems. This is key to ensure that these technologies are used effectively and responsibly.

Ultimately, sustainable farming practices require a multi-faceted approach that considers environmental, economic, and social factors. Genetic modification, when implemented responsibly and integrated into holistic farming systems, can play a crucial role in achieving this goal. By fostering environmentally friendly farming techniques and promoting healthier ecosystems, GM crops contribute to sustainable food production and long-term food security. It is essential to acknowledge that sustainable agriculture is not a singular solution but rather a combination of different practices, and genetic modification is an important piece of this puzzle.

5.1 Reduced Pesticide Use

Developing pest-resistant crops through genetic modification offers a powerful strategy to significantly reduce pesticide use in agriculture. This not only minimizes environmental harm but also contributes to a healthier and safer food system. By directly targeting the insects and pathogens that damage crops, genetically modified (GM) plants reduce the reliance on broad-spectrum pesticides that can harm beneficial insects, contaminate water sources, and accumulate in the food chain. This approach represents a substantial shift towards sustainable agriculture.

The impact of reducing pesticide use extends beyond the immediate environmental benefits. Decreased pesticide application translates to lower costs for farmers, improving their profitability and economic stability. This economic advantage is particularly beneficial for smallholder farmers in developing countries, where pesticide costs can pose a significant financial burden. Moreover, reducing pesticide use enhances the health and safety of farm workers, who are often exposed to harmful chemicals during application. Fewer pesticide-related illnesses and injuries lead to a healthier workforce and reduced healthcare costs.

Moreover, the development of pest-resistant GM crops has been shown to effectively decrease the need for pesticide application. Research has demonstrated that some GM crops exhibiting resistance to specific pests have shown a significant reduction in pesticide use, compared to conventional farming practices. In some cases, the reduction has been substantial, often exceeding 50% or even reaching 100%, depending on the pest and the specific GM crop. These studies have contributed to the growing body of evidence supporting the integration of GM technology into integrated pest management (IPM) strategies, further emphasizing the potential for sustainable agriculture.

However, the successful implementation of pest-resistant GM crops requires careful consideration of various factors. The choice of specific GM varieties must align with regional pest pressures and cropping systems. Furthermore, monitoring the evolution of pest resistance to GM crops is crucial to ensure the long-term effectiveness of this approach. Continuous research and development of new GM varieties with enhanced resistance traits are essential to maintain the sustainability of this technology. The potential for cross-pollination with wild relatives and the implications for biodiversity need to be carefully evaluated and managed to avoid unintended consequences.

Beyond direct pest resistance, the benefits extend to the overall farming system. Reduced pesticide use can positively influence soil health and biodiversity by minimizing the disruption of natural ecosystems. This indirectly contributes to improved soil fertility and water retention, leading to greater overall crop productivity and resilience. Furthermore, reduced pesticide use enhances the resilience of beneficial insects, such as pollinators, which are vital for crop production and ecosystem health. A healthy ecosystem contributes to a more stable and productive agricultural system.

Furthermore, the reduction in pesticide use can contribute to increased consumer confidence and improved food safety. Lower pesticide residues in food products can result in a safer and healthier food supply for consumers. This aligns with the increasing demand for sustainably produced food and contributes to a more robust and trustworthy agricultural sector. The decreased environmental impact also resonates with consumer preferences for eco-friendly products, potentially boosting market demand for GM crops with enhanced pest resistance.

Finally, the integration of pest-resistant GM crops into sustainable agricultural practices requires a holistic approach. This includes farmer education and training to ensure the appropriate adoption and management of these technologies. Equitable access to GM seeds and technology is crucial, particularly for smallholder farmers in developing countries. Effective regulatory frameworks are also needed to oversee the development, release, and commercialization of GM crops, ensuring their safe and responsible use while promoting agricultural innovation and sustainability. Only through a comprehensive approach can we unlock the full potential of pest-resistant GM crops to reduce pesticide use, enhance environmental sustainability, and improve food security for all. The future of agriculture lies in finding innovative and effective ways to mitigate the negative impacts of conventional farming practices, and genetically modified crops play an increasingly vital role in this endeavor. The successful integration of GM technologies requires careful consideration of the interplay between technology, environment, and societal factors.

5.2 Water Conservation

Water scarcity significantly impacts agricultural productivity. Traditional farming methods often lead to excessive water consumption, particularly in arid and semi-arid regions where many staple crops are cultivated. This unsustainable practice threatens food security, especially given the increasing global population and the predicted impacts of climate change. Developing water-efficient crops, therefore, represents a critical step towards sustainable agriculture and securing future food supplies.

Genetic engineering offers promising solutions. Scientists are actively exploring various approaches to enhance the water-use efficiency (WUE) of crops. These techniques focus on improving the plant's ability to absorb, utilize, and retain water under drought conditions. One promising method involves modifying genes that regulate stomatal function—the tiny pores on leaves that control water vapor loss during transpiration. Engineering plants with fewer, smaller, or more responsive stomata can significantly reduce water loss without compromising photosynthesis.

Another strategy centers on manipulating the plant's root system. Deeper and more extensive root systems can access deeper water sources, improving drought tolerance. Genetic modification can enhance root development and architecture, optimizing water uptake efficiency. This includes altering genes related to root growth, branching, and the production of root hairs, which increase surface area for water absorption.

Furthermore, research is ongoing to modify the genes responsible for stress response pathways. These pathways are activated under water deficit conditions, initiating physiological changes that help the plant survive drought. Genetic engineering can enhance the efficacy of these pathways, making the plants more resilient to water stress. Understanding and manipulating these intricate genetic mechanisms are central to developing drought-tolerant varieties.

The development and deployment of water-efficient crops through genetic engineering is not without its challenges. Extensive field testing is required to validate the effectiveness of these modifications under various environmental conditions and across different soil types. Furthermore, addressing potential unintended consequences, like alterations in plant growth or interactions with other organisms, is crucial. Thorough risk assessments and robust regulatory frameworks are vital to ensure the safety and sustainability of these genetically modified crops. However, the potential benefits of water-efficient GM crops are substantial. Reduced irrigation requirements can lead to significant water savings, freeing up water resources for other uses. This is particularly important in regions facing severe water shortages. Decreased water consumption also contributes to reduced energy consumption associated with irrigation, promoting environmentally friendly agricultural practices. Ultimately, the successful implementation of water-efficient crops will contribute significantly to enhanced food security in a world facing increasing water stress. This innovative approach represents a critical tool in ensuring food availability for a growing global population.

Beyond the immediate gains in water conservation, these advances also contribute to the broader sustainability goals of agriculture. Less water usage translates directly to lower environmental impact, reducing the strain on natural water resources and mitigating the consequences of water scarcity. This aligns perfectly with broader efforts to develop sustainable and resilient agricultural systems. Furthermore, the reduced need for irrigation can decrease the reliance on fossil fuel-powered pumps, thus contributing to a reduction in greenhouse gas emissions.

The development and widespread adoption of water-efficient GM crops are crucial for building a more secure and sustainable food system. They offer a powerful technological solution to a pressing global challenge—the growing scarcity of water resources. Ongoing research and careful implementation are paramount to ensuring that these innovations fulfill their potential to enhance food security while minimizing environmental impact. We must continue to invest in research, develop robust regulatory frameworks, and engage in informed public discourse to ensure the responsible development and adoption of these technologies. The future of food security is intrinsically linked to responsible water management, and genetically modified crops capable of thriving in water-scarce environments represent a promising step towards that future. A future where food production is not only efficient but also sustainable and environmentally responsible, ensuring the well-being of both current and future generations.

6 Global Food Security

The world faces a looming crisis: feeding a rapidly growing population with dwindling resources. Climate change, water scarcity, and pest infestations already threaten agricultural yields, and the problem will only intensify. This isn't just a humanitarian issue; it's a global security concern. Unstable food supplies fuel conflict and migration, destabilizing entire regions. Genetic gastronomy offers a potential solution, a way to enhance crop production and resilience, ultimately contributing to global food security.

Genetically modified crops, tailored to specific environments and challenges, can dramatically boost yields. Imagine drought-resistant maize thriving in arid regions, or rice varieties enriched with essential vitamins combating malnutrition. These are not science fiction; they're ongoing research projects, demonstrating the power of genetic engineering to address the critical needs of millions. The implications for developing nations, often hit hardest by food insecurity, are particularly profound.

Improving food access and affordability requires a multifaceted approach. Simply increasing production isn't enough; distribution networks and market structures must also be addressed. However, higher yields due to GM crops can lower production costs, making food more accessible to lower-income populations. This cost reduction can ripple through economies, reducing poverty and improving overall nutrition levels. The potential economic benefits, therefore, are immense, far outweighing potential short-term costs associated with developing and implementing genetically modified crop solutions.

Addressing concerns about the ethical implications and potential environmental impact of GM crops is crucial for their successful and responsible implementation. Open dialogue and transparent research are essential to building trust and fostering responsible development. Stringent regulatory frameworks, coupled with robust monitoring systems, can mitigate risks and ensure that genetic modifications are used safely and sustainably. We need a global commitment to responsible innovation, balancing the potential benefits with the need for careful oversight.

The role of education and public awareness is often underestimated but absolutely crucial in this context. Misinformation and unfounded fears surrounding GMOs can hamper the widespread adoption of these critical technologies. Transparent communication about the benefits, risks, and rigorous testing processes involved in developing GM crops is vital in dispelling these myths and fostering public acceptance. This means actively engaging with communities, educating consumers about the science, and addressing their legitimate concerns in an open and respectful manner.

However, the path to global food security through genetic gastronomy is not without its challenges. Regulatory hurdles, public perception, and the need for equitable access to technology remain significant obstacles. International cooperation is key, fostering collaboration between researchers, policymakers, and farmers across nations to ensure that these advancements benefit all, and that the technology reaches those who need it most. We must focus on building resilient agricultural systems that can withstand environmental stress and deliver nutritionally-rich food for all people.

Furthermore, the future of genetic gastronomy extends beyond increasing crop yields. It encompasses the development of crops with enhanced nutritional profiles, tailored to specific dietary needs and preferences. Imagine staple crops fortified with vitamins and minerals, effectively addressing micronutrient defi-

ciencies that plague many parts of the world. These developments hold the potential to revolutionize public health and contribute to a more equitable distribution of essential nutrients.
Finally, genetic gastronomy is not a silver bullet; it's one crucial tool in a broader arsenal aimed at achieving global food security. Sustainable farming practices, efficient water management, and responsible land use are all essential components. However, genetic engineering offers unique capabilities to enhance the resilience and productivity of our food systems in the face of climate change and resource scarcity. This is not just about feeding the world; it is about building a more sustainable, just, and secure future for all.

6.1 Addressing Food Shortages

The world faces a looming crisis: food shortages threaten billions. Climate change, dwindling resources, and escalating populations create a perfect storm. Addressing this requires a multifaceted approach, and genetic modification offers a powerful tool in our arsenal. It isn't a silver bullet, but a significant step forward. We must recognize the potential of genetically modified crops to revolutionize agriculture.
Current agricultural practices often prove insufficient against the

pressures of a changing climate. Traditional crops struggle with drought, pests, and diseases, leading to significant yield losses. Genetically modified crops, however, can be engineered for increased resilience, allowing them to thrive even under stressful conditions. Imagine drought-resistant maize thriving in arid regions, or rice varieties fortified with essential vitamins growing in nutrient-poor soils. This translates directly into more food for more people.

Developing nations bear the brunt of food insecurity. Poverty, limited resources, and inadequate infrastructure exacerbate the problem. Genetically modified crops offer a solution, allowing farmers to cultivate higher yields with less input. Pest-resistant crops reduce the need for harmful pesticides, protecting both the environment and the health of farmworkers. Higher yields mean increased income for farmers and greater food availability for their communities. The impact extends beyond mere subsistence; improved nutrition can lead to improved health outcomes and contribute to economic development.

However, widespread adoption faces numerous challenges. Concerns about the environmental impact of genetically modified crops and their potential long-term effects necessitate rigorous scientific research and transparent regulation. Public perception, influenced by misinformation and fear-mongering, also plays a crucial role. Education and open dialogue are key to addressing these concerns, fostering trust and understanding around the benefits of genetic modification. This involves clear communication about the safety protocols and rigorous testing processes that underpin the development and release of GM crops.

Furthermore, equitable access to genetically modified crops and the technology behind them is paramount. Intellectual property rights and licensing agreements must be carefully structured to ensure that developing nations benefit from these advancements without being exploited. International cooperation and collaboration are crucial to share knowledge, resources, and best practices, enabling the widespread adoption of these technologies in a fair and responsible manner. This requires a commitment to fostering agricultural innovation in the global south, equipping farmers with the knowledge and tools they need to succeed.

The future of food security hinges on innovation and collaboration. Genetic modification is not a solution in isolation; it must be part of a broader strategy that integrates sustainable agricultural practices, improved infrastructure, and equitable distribution systems. By combining genetic engineering with improved farming techniques, we can maximize the positive impact, ensuring food security for present and future generations. This holistic approach is essential for achieving sustainable and equitable outcomes.

Beyond addressing immediate needs, we must anticipate future challenges. Climate change continues to exert immense pressure on our food systems. Genetic modification offers powerful tools to adapt to these changes, allowing us to cultivate crops that are resilient to drought, heat, and other environmental stressors. By creating more robust and adaptable crops, we can secure a more stable food supply and mitigate the effects of climate change on food production. This resilience is key to safeguarding global food security in the face of an uncertain future. The capacity to adapt and innovate is critical to sustaining our food systems.

Investing in research and development is critical for the advancement of genetic gastronomy. New technologies and techniques continually emerge, offering the potential to enhance crop yields, improve nutritional value, and enhance crop resilience. This requires sustained financial investment and international collaboration, promoting shared knowledge and innovation across borders. Continuous improvement and adaptation are essential for addressing the ever-evolving challenges of food production. This is an ongoing process requiring consistent dedication to research and collaboration.

Ultimately, addressing food shortages necessitates a commitment to sustainable and ethical practices. Genetic modification, when used responsibly and coupled with other innovative strategies, provides a potent tool to combat food insecurity and build a more resilient food system. A future where hunger is eradicated requires both technological innovation and a commitment to social justice. This collective effort is essential for ensuring food security for all.

6.2 Impact on Developing Nations

Developing nations face unique challenges regarding food security, often grappling with malnutrition, limited resources, and climate change vulnerabilities. Genetically modified (GM) crops present a powerful tool to address these issues, offering the potential for enhanced nutrition, increased yields, and improved resilience to environmental stressors. However, the successful implementation of GM technology in these contexts requires careful consideration of several critical factors.

The nutritional deficiencies prevalent in many developing countries can be directly tackled through biofortification. GM crops enriched with essential vitamins and minerals like Vitamin A, iron, and zinc can significantly improve public health outcomes, reducing the burden of micronutrient malnutrition, a significant contributor to child mortality and impaired cognitive development. The success of Golden Rice, engineered to produce beta-carotene, a precursor to Vitamin A, exemplifies the potential of this approach. Furthermore, the ability to tailor GM crops to specific regional nutritional needs allows for highly targeted interventions.

Beyond enhanced nutrition, GM crops offer the potential to significantly increase crop yields. Developing nations often rely on rain-fed agriculture, making them highly susceptible to drought. Drought-tolerant GM crops can significantly reduce yield losses during periods of water scarcity, ensuring more stable food production even in the face of unpredictable weather patterns. Similarly, pest-resistant varieties can reduce reliance on costly and potentially harmful pesticides, making farming more economically viable for smallholder farmers who often lack access to resources or training in chemical pest management.

However, the widespread adoption of GM crops in developing nations requires more than just technological advancements. Access to appropriate technology, seeds, and training is crucial for successful implementation. Many smallholder farmers lack the resources to purchase GM seeds, highlighting the need for affordable seed distribution systems and public-private partnerships that ensure equitable access. Furthermore, effective farmer education programs are essential to address concerns, provide technical assistance, and build confidence in the use of GM technology. It is vital to provide training on how to plant, manage, and harvest these crops effectively.

The regulatory landscape surrounding GM crops is another critical aspect. Stringent, yet transparent and accessible regulatory frameworks are needed to ensure the safety and responsible use of GM technology while avoiding overly burdensome regulations that can hinder innovation and access. Collaboration between international organizations, governments, and local communities is crucial to developing appropriate regulations that promote food security while addressing potential risks. Effective communication and transparency are vital to ensuring public trust and acceptance. This often involves addressing common misconceptions about GM foods, emphasizing the rigorous safety testing involved, and highlighting the benefits for both farmers and consumers.

Finally, the economic impact of GM crops on developing nations requires careful analysis. While increased yields and reduced input costs can significantly benefit farmers, it's essential to consider the potential effects on market prices and the livelihoods of those dependent on traditional farming practices. Careful market studies, economic impact assessments, and appropriate support mechanisms are vital to avoid unintended negative consequences. A nuanced approach, that acknowledges potential challenges while emphasizing the benefits, is crucial for a successful and sustainable implementation of GM crops in developing nations. This includes providing economic incentives to farmers, ensuring fair market access for their produce, and supporting the development of value-added processing industries to maximize economic returns. Only with a holistic approach that integrates technological innovation with social, economic, and regulatory considerations can genetic gastronomy realize its full potential in improving food security and nutrition in developing nations.

6.3 Food Access & Affordability

Access to nutritious and affordable food remains a significant global challenge, directly impacting health and societal well-being. Genetically modified (GM) crops offer a powerful tool in addressing this multifaceted issue, but their potential is only fully realized when considering the broader context of food systems. Simply producing abundant, nutritious food isn't enough; that food needs to reach those who need it most.

The affordability of GM crops is directly linked to production costs. Lower production costs, stemming from reduced pesticide application or improved water-use efficiency in genetically modified varieties, can translate to lower market prices. This is especially relevant in developing nations where farmers often operate on tight margins. However, factors beyond the farm gate, such as processing, transportation, and retail markups, all influence final consumer prices. Therefore, policies promoting fair trade practices and transparent supply chains are crucial. Research into cost-effective processing and storage methods for GM crops is also vital for ensuring affordability in regions with limited infrastructure.

Furthermore, access transcends affordability. Geographic location, inadequate infrastructure, and societal disparities can create "food deserts," areas where nutritious food is scarce. GM crops alone won't solve this. Targeted interventions are needed, including investments in efficient transportation networks, community gardens, and initiatives that directly address food insecurity at a local level. Policies focusing on market access for smallholder farmers—who often cultivate GM crops—are also essential, enabling them to participate more effectively in food distribution networks. Indeed, empowering these farmers economically makes them essential stakeholders in a sustainable and accessible food system.

Innovative distribution strategies are also essential. Mobile markets, community-supported agriculture (CSA) schemes, and the expansion of food banks can bridge gaps in accessibility. These approaches require collaboration between government, non-governmental organizations, and private businesses. Technological advancements, such as precision agriculture using data-driven optimization of resource allocation, could also play a crucial role in enhancing efficiency and reducing waste in the food supply chain. This enhances both access and affordability for marginalized communities.

The societal impact of improved food access and affordability through GM crops is substantial. Enhanced nutrition can lead to better health outcomes, particularly for vulnerable populations. Reduced food insecurity can increase school attendance rates among children, allowing them to focus on education rather than hunger. Increased farm incomes for smallholder farmers can contribute to economic growth in rural areas. Ultimately, the responsible development and implementation of GM crops can be a significant contributor to a more just and equitable food system worldwide. However, it's crucial to acknowledge that GM technology is merely one piece of a larger puzzle, requiring concerted effort across multiple sectors to fully unlock its potential for improving food access and affordability on a global scale.

The goal is not merely increased production, but a systematic overhaul of food systems to ensure access for all.

7 The Future of Genetic Gastronomy

Gene editing technologies are rapidly advancing, promising even more precise and efficient modifications than currently available. We're moving beyond simply enhancing existing traits; researchers are exploring the creation of entirely novel traits, potentially leading to crops with enhanced flavors, textures, and even entirely new nutritional profiles unimaginable today. Imagine a tomato bursting with antioxidants far surpassing anything found in nature, or a rice grain fortified with all the essential vitamins and minerals a person needs. These aren't merely futuristic fantasies; they're within the realm of possibility.

The regulatory landscape surrounding genetically modified organisms (GMOs) is continuously evolving. Different countries have diverse regulations and approval processes, leading to complexities in international trade and access to these advanced technologies. A more harmonized global regulatory framework, based on robust scientific evidence and risk assessment, is crucial to facilitating the widespread adoption of beneficial GMOs and avoiding trade barriers that could hinder progress, especially for developing nations that need these technologies most. Transparency and public engagement in the regulatory process are essential for building trust and ensuring equitable access.

Consumer perception remains a critical factor influencing the success of genetic gastronomy. Many consumers harbor misconceptions or concerns about GMOs, fueled by misinformation and a lack of understanding of the science behind these technologies. Effective communication strategies are therefore needed to address these concerns, emphasizing the benefits of GMOs in terms of nutrition, sustainability, and food security. Education, transparency, and clear labeling are vital in building consumer confidence and fostering acceptance. Open dialogue between scientists, policymakers, and the public can help dispel myths and build a better understanding of this complex field, ensuring that the potential of genetic gastronomy is fully realized. The future of food security may well hinge on overcoming these communication hurdles.

Beyond specific traits, we're also seeing the application of genetic modification to improve the entire agricultural system. For example, scientists are working on crops that are more resilient to pests and diseases, reducing the need for pesticides and improving yield stability. This has clear environmental benefits, contributing to sustainable farming practices and mitigating the impact of climate change. Likewise, research is underway to create crops that are more water-efficient, critical in regions facing water scarcity. Genetic gastronomy is not just about creating better food; it's about creating a more sustainable and resilient food system.

While enhancing nutritional content and improving crop resilience have been the primary focus thus far, the future holds opportunities to explore more innovative applications. Scientists are investigating the possibility of genetically modifying crops to improve their taste and texture, making them more appealing to consumers. Imagine fruits and vegetables that are sweeter, juicier, and more flavorful, even after they've been stored or transported. Moreover, genetic modification could help create crops that are better suited to specific climates and soil conditions, allowing for cultivation in regions currently unsuitable for particular crops. The expansion of cultivatable land and the ability to grow nutritious crops in previously barren areas are promising avenues to explore. This would impact global food security greatly and would help reduce dependency on specific regions for certain types of food.

Finally, the integration of emerging technologies, such as advanced data analytics and artificial intelligence (although we cannot talk specifically about it), will play a crucial role in shaping the future of genetic gastronomy. These tools can aid in the design and optimization of genetic modifications, leading to more precise and effective outcomes. Data analysis can also help predict and mitigate risks associated with GMOs, while AI can be used for the modeling of complex interactions within the food system and for the precision design of crops. A sophisticated approach which includes all these factors will be essential in order to guarantee a sustainable and impactful food-related future for everybody, everywhere. The future of genetic gastronomy promises to revolutionize not only how we produce food, but how we feed a growing global population sustainably and nutritiously. The convergence of these technologies offers a path toward a future of greater food security and improved public health. It's a future that depends on continued research, responsible regulation, and open communication with the public.

7.1 Emerging Technologies

Gene editing is rapidly evolving, moving beyond single-gene modifications. Multiplex gene editing allows for simultaneous changes across multiple genes, offering unprecedented control over crop traits. Imagine engineering a tomato with enhanced flavor, improved shelf life, and increased disease resistance – all achieved in a single modification process. This technology holds immense promise for creating crops perfectly tailored to specific needs, maximizing both yield and nutritional value. The complexity, however, necessitates rigorous safety assessments and thorough risk analyses before widespread implementation.

Next-generation sequencing technologies are revolutionizing our understanding of plant genomes. High-throughput sequencing provides detailed genetic information, enabling us to identify genes responsible for desirable traits with greater accuracy and speed than ever before. This allows for more targeted gene editing, minimizing unintended consequences. We can now dissect complex traits, identifying the intricate interplay of multiple genes, rather than relying on a reductionist approach that targets single genes in isolation. This deeper understanding translates to more effective and predictable outcomes in crop improvement. It's a paradigm shift from guesswork to precision engineering.

Artificial intelligence and machine learning are increasingly integrated into the design and development of GM crops. AI-powered algorithms can analyze vast datasets of genomic information, identifying genes associated with desired traits with improved accuracy and efficiency. These algorithms can predict the effects of genetic modifications, minimizing the need for lengthy and expensive trial-and-error experiments. This accelerates the entire process of crop improvement, potentially leading to faster adaptation to the changing climate and more robust food production systems. We're seeing more and more applications where AI streamlines the laborious tasks of genetic modification, freeing up researchers to focus on innovation.

Synthetic biology offers exciting possibilities for creating entirely new metabolic pathways within plants. Imagine designing plants that produce entirely new nutrients or compounds, enhancing their nutritional profile beyond what's naturally possible. We could engineer crops to synthesize essential vitamins, produce pharmaceuticals, or even generate biofuels. While still in its early stages, this technology has the potential to reshape agriculture and create crops with functionalities far exceeding anything we can achieve through traditional breeding or gene editing alone. The ethical implications of such power require careful consideration, ensuring responsible development and deployment.

While gene editing techniques hold much promise, concerns regarding unintended consequences persist. Off-target effects, where modifications occur in unintended locations within the genome, are a major focus of ongoing research. Advancements in CRISPR-Cas9 technology and the development of more precise gene editing tools are helping mitigate these risks. Researchers are developing methods to detect and minimize off-target effects, ensuring the safety and predictability of the genetic modifications. Robust regulatory frameworks are essential to govern this technology's use and ensure responsible innovation. We need to balance the potential benefits with the need to safeguard the environment and human health.

The integration of these emerging technologies promises a future where food security is strengthened, nutritional deficiencies are minimized, and agriculture becomes more sustainable. However, we must prioritize robust safety assessments and responsible innovation. This necessitates open dialogue among scientists, policymakers, and the public to ensure the ethical and equitable application of these advancements, creating a future where genetically modified foods contribute to a healthier planet and a more food-secure world for all. The challenges are significant, but the potential rewards are even greater. It's our collective responsibility to navigate this technological frontier wisely.

7.2 Regulatory Landscape

Navigating the complex world of genetically modified (GM) food requires understanding the regulatory landscape. This isn't a simple set of rules; it's a constantly evolving interplay of national and international laws, agency interpretations, and public opinion, profoundly shaping the development and commercialization of GM crops. Each country establishes its own regulatory framework, leading to significant variations in approval processes, labeling requirements, and permissible GM traits. These differences can create challenges for international trade and research collaborations.

For instance, the European Union's regulatory system is notably rigorous, involving extensive risk assessments and demanding comprehensive data on the safety and environmental impact of GM crops before approval. This stringent approach contrasts sharply with the more streamlined processes often seen in other regions such as the United States, where the regulatory burden is comparatively lighter, fostering quicker adoption of new GM varieties. Such contrasting approaches underscore the importance of navigating these complexities if you intend to work internationally. Understanding this framework is crucial for ensuring compliance and success in the field.

The approval processes themselves are multifaceted. They usually involve multiple stages, beginning with initial assessments of potential risks to human health and the environment. Safety testing encompasses a wide range of parameters, from toxicological studies evaluating potential negative impacts on human and animal health to environmental assessments focusing on potential impacts on biodiversity and non-target organisms. Subsequently, regulatory agencies conduct rigorous analyses of the data presented by developers, often taking months or even years to reach a decision. This is a time-consuming process often resulting in significant financial and logistical hurdles for companies developing new GM crops. The length of the process can deter smaller businesses.

Furthermore, labeling regulations vary considerably worldwide. Some nations require mandatory labeling of GM products, informing consumers about the presence of genetically modified ingredients. Other countries have adopted voluntary labeling schemes or have no labeling requirements at all. These varying mandates directly influence consumer choices and market dynamics, highlighting the significance of keeping abreast of the latest labeling regulations in your area of operation. Keeping track of this is difficult; the landscape is constantly evolving.

Beyond national regulations, international organizations, such as the Codex Alimentarius Commission, play a pivotal role in harmonizing food safety standards worldwide. This commission develops guidelines and recommendations for the assessment and management of risks associated with GM foods, aiming to provide a basis for consistent regulatory approaches across nations. However, even with such guidelines, significant differences in regulatory approaches persist, reflecting varying risk perceptions and socio-political contexts. This is why global collaboration is vital to create universally acceptable standards.

Beyond the technicalities of approval processes and labeling, public perception and political considerations significantly influence the regulatory landscape. Public opinion regarding GM foods can be highly variable, and influencing factors such as media coverage, scientific literacy, and consumer trust have a profound impact on policy decisions. These influences often translate into either supporting or opposing regulations that govern the development and use of GM foods. Public opinion needs to be considered.

Ultimately, successfully navigating this complex regulatory environment necessitates proactive engagement with relevant agencies, thorough understanding of national and international regulations, and continuous monitoring of emerging scientific findings and policy developments. Staying abreast of the ever-shifting regulatory landscape is vital to fostering innovation in genetic gastronomy while ensuring responsible and sustainable agricultural practices. Understanding these regulations is a crucial part of responsible agricultural innovation. Staying aware of these changes is a constant necessity.

7.3 Consumer Perception

Consumer attitudes towards genetically modified (GM) foods are complex and often influenced by factors beyond scientific evidence. Fear of the unknown, coupled with misinformation campaigns, has fostered skepticism and, in some cases, outright opposition. This distrust, often fueled by emotional responses rather than rational analysis, presents a significant hurdle to the widespread adoption of GM crops, despite their potential benefits for food security and sustainability. Overcoming this challenge requires a multifaceted approach that prioritizes transparency, education, and open dialogue.

Public perception is heavily shaped by media portrayals, which can be sensationalized and lacking in nuance. Negative stories, often focusing on hypothetical risks, gain disproportionate attention compared to positive reports on the real-world benefits of GM technologies. This creates an unbalanced narrative that reinforces existing anxieties. Consequently, many consumers remain unaware of the rigorous safety testing and regulatory oversight that GM crops undergo before reaching the market. Addressing this requires a proactive strategy of disseminating accurate information through various channels, including social media and community outreach programs.

Building trust is paramount. Scientists and policymakers must engage with the public directly, addressing concerns honestly and transparently. This means actively participating in community discussions, responding to questions and criticisms openly, and demonstrating a genuine commitment to ethical and responsible innovation. Clear communication about the scientific process, the risk assessment protocols used for GM foods, and the potential benefits for both human health and the environment can help alleviate anxieties and build confidence.

Transparency in labeling is another crucial element. Clear and accessible labeling that accurately reflects the genetic modification process can empower consumers to make informed choices. However, overly alarmist or misleading labels can further fuel anxieties and distrust. A balanced approach that provides accurate information without resorting to fear-mongering is essential. Moreover, consistent labeling across different regions and jurisdictions can ensure a unified and transparent system that fosters consumer trust.

Finally, fostering a collaborative environment between researchers, policymakers, industry stakeholders, and the public is essential. Open dialogue, engaging in constructive discussions, and actively listening to consumer concerns are key to building a shared understanding and addressing anxieties. By working together to create a climate of transparency and mutual respect, we can pave the way for a more informed and accepting public perception of GM crops and their potential to improve our food systems. Ignoring or dismissing public concerns will only strengthen resistance to technological progress. A more collaborative and communicative approach is needed to ensure that consumers fully comprehend the long-term benefits of genetically modified foods. The future of genetic gastronomy hinges on bridging this gap between scientific innovation and public understanding, and ensuring the acceptance of a technology that promises significant improvements to human health and environmental sustainability. The future demands a holistic strategy that combines cutting-edge science with effective communication and a commitment to transparent dialogue. Only then can we harness the full potential of genetic gastronomy to create a more sustainable and secure global food supply.